America's Increasing Reliance on Natural Gas: Benefits and Risks of a Methane Economy

I0484833

Report of the Critical Issues Forum
November 19-20, 2014, Fort Worth, Texas

Written by Timothy Oleson, Ph.D.

Contents

© 2015 American Geosciences Institute
isbn: 978-1-508843-50-4
American Geosciences Institute
4220 King Street
Alexandria, VA 22302-1507 U.S.A.
Phone: +1 (703) 379-2480
Fax: +1 (703) 379-7563
agi@americangeosciences.org
www.americangeosciences.org
For more information on Critical Issues Forums, go to
www.americangeosciences.org/policy/ci-forums.

Design by Brenna Tobler, AGI
Cover background image © Sergey Nivens/Shutterstock.com, pipeline image © DabartiCGI/Shutterstock.com

american
geosciences
institute

connecting earth, science, and people

Critical Issues Forum Planning Committee

Scott Cameron, Consulting geologist
Nick Tew, Geological Survey of Alabama and State Oil and Gas Board of Alabama
Scott Anderson, Environmental Defense Fund
Jesse Ausubel, The Rockefeller University

Forum Agenda, November 19 and 20, 2014

Wednesday, November 19

8:00 - 8:30	Registration
8:30 - 8:45	Introduction & Welcoming Remarks
	Eric Riggs, President, AGI
	P. Patrick Leahy, Executive Director, AGI
8:45 - 10:00	Keynote Address
	Scott W. Tinker, Director, Bureau of Economic Geology, The University of Texas at Austin, and State Geologist of Texas
10:00 - 10:30	Morning Break
10:30 - 12:15	**Session 1: Outlook for natural gas supply**
	John B. Curtis, Colorado School of Mines
	Richard Nehring, Nehring Associates
	L. Renee Orr, Bureau of Ocean Energy Management
	David Pursell, Tudor, Pickering, Holt & Co.
	Wendy Harrison, Colorado School of Mines (moderator)
12:15 - 1:15	Lunch
1:15 - 2:45	**Session 2: Forecasts of natural gas demand**
	Eyal Aronoff, Fuel Freedom Foundation
	David Levinson, University of Minnesota
	Kenneth Medlock, Rice University
	Jesse Ausubel, Rockefeller University (moderator)

2:45 - 3:15	Afternoon Break
3:15 - 4:45	**Session 3: Environmental, health, and safety impacts**
	Mark Brownstein, Environmental Defense Fund
	Doug Jordan, Southwestern Energy Company
	Alan Krupnick, Resources for the Future
	Richard Liroff, Investor Environmental Health Network (moderator)
5:00 - 6:30	Reception
6:30 - 8:00	Dinner and Evening Keynote Address
	Katherine Lorenz, President and Treasurer, The Cynthia & George Mitchell Foundation

Thursday, November 20

8:00 - 8:30	Light Breakfast
8:30 - 10:00	**Session 4: Drivers of and barriers to natural gas development in North America**
	Lawrence Bengal, Arkansas Oil and Gas Commission
	Kitty Milliken, Bureau of Economic Geology, University of Texas
	Randy Randolph, Southern Gas Association
	Berry H. (Nick) Tew, Jr., Geological Survey of Alabama and State Oil and Gas Board of Alabama (moderator)
10:00 - 10:30	Morning Break
10:30 - 12:00	Facilitated wrap-up discussion

The American Geosciences Institute (AGI)

serves as a voice of shared interests in the profession and plays an active role in increasing public awareness of the vital role the geosciences play in society. We created the Critical Issues Forum series as a platform to reach a broader audience of decision makers, including those at the regional, state, and local levels, and to improve public understanding and perception of the geosciences.

I am pleased to present this report summarizing the stimulating presentations and discussions from the inaugural AGI Critical Issues Forum. The two-day meeting examined many dimensions of *America's Increasing Reliance on Natural Gas: Benefits and Risks of a Methane Economy* and considered two major questions:

- Is a natural gas-dominant economy achievable in North America?
- Would a natural gas-dominant economy be desirable?

We extend our thanks to all who participated in the Forum and we look forward to hosting other vital conversations highlighting the importance of the geosciences in society.

Warm regards,

Dr. P. Patrick Leahy
Executive Director
American Geosciences Institute

AGI thanks the following organizations for their support of the Critical Issues Forum.

In the U.S., we don't really understand much about energy: where it comes from, the scale of the demand, or the benefits and challenges of producing different kinds of energy. Energy has lifted much of the world out of poverty and is fundamental to improving the health and growth of both developed and developing economies. But nothing is perfect in the energy world, and there is no one-size-fits-all solution for all countries and communities.

I don't know where things will stand 50 years from now, but I do know that, like today, we are still going to be looking for sources that are affordable, accessible, reliable and sustainable. Those tenets will drive the energy mix, whatever it turns out to be.

The challenge is that addressing energy issues in a reasonable way requires many players, each with their own knowledge and understanding, to leave their respective corners: national, state and local governments; regulatory agencies; the energy production and distribution industries, as well as industries that consume large amounts of energy; academics; and nongovernmental organizations.

I call that space — where reasonable people from different groups are willing to come together to compromise — the **radical middle**. You need not give up your viewpoint, you just need to be willing to see other sides and try to find common ground on critical issues that everybody can work toward achieving.

— Scott Tinker, Director of the Bureau of Economic Geology at the University of Texas at Austin, and State Geologist of Texas

America's Increasing Reliance on Natural Gas: Benefits and Risks of a Methane Economy
Report of the Critical Issues Forum

Every day, most Americans work, study or relax in climate-controlled, well-lit rooms. Every day, most Americans make phone calls or send emails, wash their hands with clean water and pull food from refrigerators. And every day, most of the 319 million Americans either ride in one of the 253 million vehicles on the road, or take a subway, train, plane, boat or some other means of transportation.

That all of these activities — integral to our daily lives — require energy won't come as news to anyone. But the amount of energy needed to fuel our way of life is often under-appreciated, and rarely do we give much thought to where the energy that drives modern life is sourced. Even with great advances in efficiency, total energy use today in the United States is twice what it was 50 years ago, and globally we trail only China, which has a population more than four times as large. On average, each American consumes the equivalent of about 2,420 gallons of oil annually — more per person than in any other of the world's 35 most-populous nations.

Visualize 2,420 gallons of oil multiplied by 319 million people, and the enormous scale of our energy requirement comes into focus. Supplying all that energy — from sources that are affordable, accessible, reliable and sustainable — is a challenge equally enormous. Since the early 20th century, abundant, energy-dense fossil fuels including petroleum, natural gas and coal have been our go-to sources to meet most demand. And that continues today, with those three fuels accounting, respectively, for 36, 27 and

© Shutterstock.com/wavebreakmedia

18 percent of all U.S. energy consumption. Of course, we use other sources as well: nuclear (9 percent of total U.S. energy usage), hydroelectric (3 percent), and renewable sources such as biomass, wind, geothermal and solar (6 percent combined).

The mix of sources we've used has long been dynamic, shifting historically — though largely out of the public eye — in response to changes in supply, demand, U.S and global economic activity, technological innovation and available infrastructure. The U.S. led the world in total petroleum

Historical Energy Consumption in the United States (1949-2013)

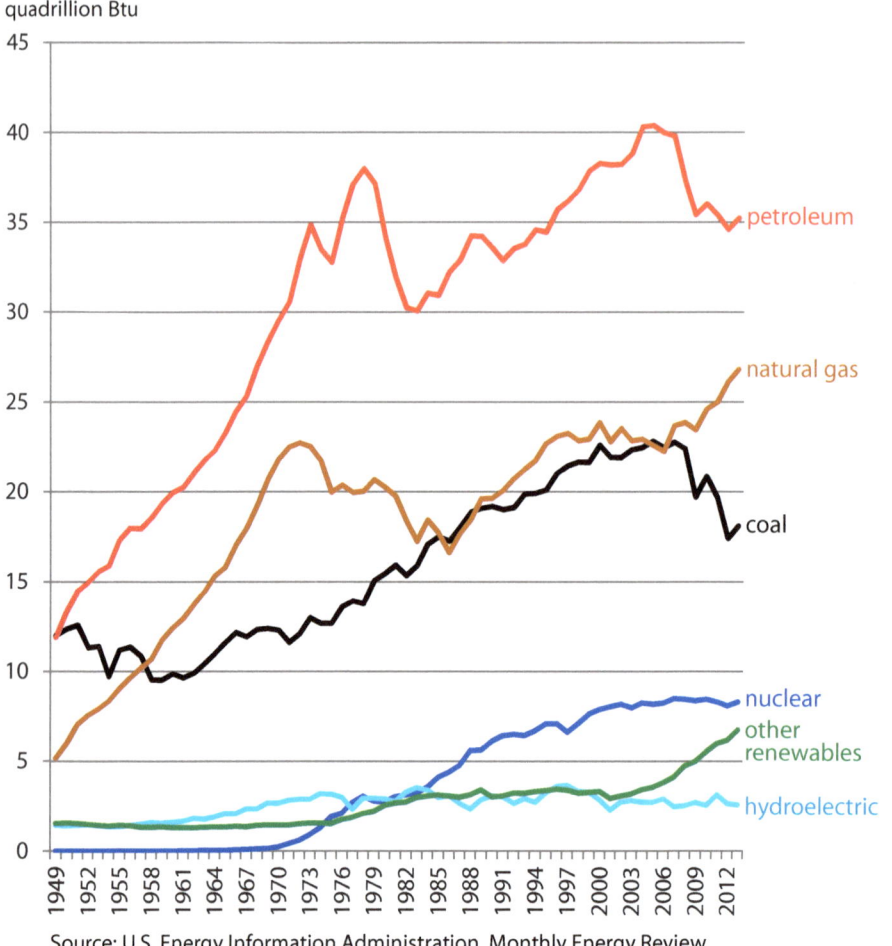

quadrillion Btu

Source: U.S. Energy Information Administration, Monthly Energy Review
Chart by Leila Gonzales, AGI

production until the 1970s, after which it was eclipsed by output from other countries. But since the mid-2000s, U.S. oil and gas production has seen a resurgence, fueled by advances in technology and geologic understanding.

The economic potential of the current oil and gas boom is undeniable for the United States, both as an engine of economic growth and as a measure of energy security. Yet, the rapid expansion of the domestic footprint

of energy development has also dramatically increased awareness of the challenges involved. And increased publicity about the potential hazards and impacts of energy production and transport has led to conversations about energy and the environment that have grown louder and more fraught with emotion, giving the impression of an issue defined by strongly entrenched positions and with little opportunity to find common, or middle, ground.

However, there is more opportunity than there might seem. Most Americans do not perceive economic and environmental prosperity as an either-or proposition — quite the opposite. According to polling by the University of Texas at Austin, for example, roughly half of respondents said that economic and environmental concerns go "hand-in-hand," whereas just one-quarter said that one or the other "should always be given priority" over the other. (See page 9.)

When it comes to energy, every source has upsides and downsides — whether due to cost, accessibility, reliability or potential as a hazard. As with all complex problems, there is no easy, single solution to solving the future energy requirements of a growing economy. But inflexibility and an

> U.S. oil and gas production has seen a resurgence, fueled by advances in technology and geologic understanding. Between 2007 and 2013, for example, U.S. shale gas production rose 300 percent.

unwillingness to consider alternative ideas is a recipe for stagnation, not success. And finding common ground — on the desire for ample affordable energy and for a continued push for the implementation of best practices in energy production and environmental protection, for example — through open and honest communication is paramount in ensuring we can maintain the high standards of living we enjoy across the country.

> A majority of the world's energy consumption is hydrocarbon-based. The projection is for that to not change very much.
>
> – Ken Medlock, Center for Energy Studies, James A. Baker III Institute for Public Policy, Rice University

Based on current and expected future energy demands in the U.S., coupled with the anticipated availability of resources and infrastructure, the U.S. will continue to rely on a diverse mix of energy sources in the coming decades. This mix will most likely still be led by fossil fuels because the technology and infrastructure for efficient generation and/or transmission of power from sources like nuclear and renewables is expected to be inadequate to fulfill the majority of our demand in the near future. And improved supply of fossil fuels, especially natural gas, will likely play a critical role in bridging the transition of the economy towards future energy sources.

Renewable energy consumption has been growing annually by about 6 percent on average since 2008, thanks

© Shutterstock.com/Pressmaster

Natural Gas for Vehicles

TURN OFF ENGINE   NO SMOKING

COMPRESSED NATURAL GAS

THIS SALE $

GASOLINE GALLON EQUIVALENT

3600 PSI Fill
FILL COMPLETION IN %

CNG
MINIMUM
90%
METHANE

PRICE PER GASOLINE GALLON EQUIVALENT

REFUELING INSTRUCTIONS
STOP ENGINES, NO SMOKING, FLAMMABLE GAS

© Shutterstock.com/bikerlondon

carbon dioxide into the atmosphere than those two fuels; and it is already in common use in many areas — meaning the base infrastructure needed to convert to a natural gas-led energy economy is already in place. Yet, natural gas has drawbacks as well: It contains less energy per volume than other fossil fuel sources, creating difficulties for transporting and storing it, as well as for its use as a transportation fuel; it still produces more carbon dioxide than renewable sources; and although some infrastructure exists to support heavy gas consumption, more would be needed in a majority-methane economy.

> A *methane economy*, as defined here, is an economy in which natural gas provides the leading share of primary energy consumption nationally.

largely to tax credits and government investments that have stimulated research and development. Sustaining that brisk pace, however, will be difficult due to the ever-increasing need for materials to build the required infrastructure, as well as the land (or sea) surface area on which to install it. And even if this pace were sustained, it would still take until mid-century for renewable consumption to equal current fossil fuel consumption. This is not an argument against development of non-fossil resources, but is instead one illustration of the challenge of transitioning away from a fossil fuel-dominated economy.

In recent years, natural gas, which is composed mostly of methane, has emerged as an appealing option to meet the majority of our energy needs — thanks in large part to new technologies that allow gas to be extracted from shale rock buried below the surface. Natural gas has particular advantages for the U.S.: There is a large domestic supply; it burns more efficiently than petroleum and coal so it releases less

For all we do know about natural gas, there is more we don't. Recognizing this and that we as individuals, states and a country must make critical decisions about where our future energy will come from — the American Geosciences Institute (AGI) convened a meeting of experts to analyze the potential for a methane-dominant economy in the U.S. This first-of-its-kind Critical Issues Forum brought geoscientists, economists and other natural gas experts from academia, industry, government and nongovernmental organizations (NGOs) together to consider issues of supply, demand and environmental health and public safety related to natural gas, as well as the barriers to and enablers of a

Attitudes Toward Natural Gas

Question: Which of the following best reflects your views on the relationship between the environment and the economy?

1) There does not need to be a trade-off between the economy and the environment — they go hand-in-hand. (47%)

2) Protecting the environment is the best way to achieve economic goals. (15%)

3) Ensuring economic growth is the best way to achieve environmental goals. (14%)

4) The environment should always be given priority over economic growth. (14%)

5) Economic growth should always be given priority over protecting the environment. (11%)

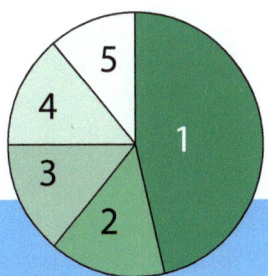

Question: To what degree do you perceive each of the following to be a benefit of domestic natural gas production?

1) Creates jobs —
   Benefit: 67%; Not a benefit: 6%;
   Neutral: 18%; Don't know: 9%

2) Lowers costs —
   Benefit: 65%; Not a benefit: 6%;
   Neutral: 19%; Don't know: 10%

3) Provides energy security —
   Benefit: 64%; Not a benefit: 5%;
   Neutral: 20%; Don't know: 11%

4) Increases energy efficiency —
   Benefit: 62%; Not a benefit: 6%;
   Neutral: 21%; Don't know: 11%

5) Boosts U.S. manufacturing —
   Benefit: 60%; Not a benefit: 6%;
   Neutral: 22%; Don't know: 12%

6) Lowers carbon emissions —
   Benefit: 53%; Not a benefit: 11%;
   Neutral: 22%; Don't know: 13%

Source: The University of Texas Austin Energy Poll, September 2014 (http://www.utenergypoll.com)

©Shutterstock.com/Carey Kalscheuer

potential methane economy. The goal was not to formulate a singular position or suggest specific policy, but rather to foster open, frank dialogue about a critically important issue and to explore how the varied interests in a methane future might best find compromise to advance our common energy goals.

This report offers an overview of the key conceptual ideas that arose in the forum and outlines the suggested approaches that different stakeholder groups can take to advance the conversation on energy in the U.S.

The Methane (R)Evolution

The U.S. is in the midst of a revolution in natural gas production; although it may be more apt to say we are in a period of ongoing evolution. Natural gas has long been produced from so-called conventional deposits where it collects in high concentrations — the first such well was dug in 1821 in New York. It is also commonly extracted as a byproduct of petroleum and coal production, during which it is often burned off, or flared, as unwanted waste.

In 1950, natural gas accounted for just 19 percent of U.S. fossil fuel production, and as recently as 2010, it still ranked behind coal in this category. Now, natural gas holds twice the stake in our fossil fuel production

that it did in 1950, putting it ahead of coal and oil. The dramatic change has been particularly rapid since the mid-2000s, due primarily to the emergence of effective technologies for horizontal drilling and hydraulic fracturing — also known as "fracing" or "fracking" — that allow gas ("shale gas") to be extracted from layers, or beds, of shale rock that have sufficient organic matter buried deep underground.

> We are in the middle of a scientific revolution in our understanding of Earth's most abundant sedimentary material: shale.
>
> – Kitty Milliken,
> Bureau of Economic Geology,
> Jackson School of Geosciences,
> University of Texas at Austin

Between 2007 and 2013, shale gas production rose 300 percent, and shale gas wells now produce more natural gas in the U.S. than any other single type of gas well. Domestically, the U.S. is endowed with a number of expansive gas-bearing shale formations that have the right mix of organic matter and maturity, some of which were almost unknown as resources even into the 2000s. Examples include the Marcellus (mostly in Pennsylvania, New York and West Virginia), Eagle Ford (Texas), Permian (Texas, New Mexico), Haynesville (in Texas and Louisiana), Barnett (in Texas, where widespread fracking was first employed) and Fayetteville (in Arkansas) shales. In addition to these shale beds, other substantial resources are either known or are predicted to exist — both onshore and offshore.

Production and Delivery

The first step in determining whether a methane economy is feasible is assessing the available quantity of natural gas. Is there enough gas to supply a majority-methane economy far enough into the future — that is, until we have the capability to source most of our energy from non-fossil fuel resources — to justify necessary long-term investments in workforce development and infrastructure for processing, storage and transportation (pipelines, refineries and ports, for example), as well as for consumption (natural gas-fueled cars and electricity generating stations, for example)?

Most experts agree that, despite having produced more than 1,000 trillion cubic feet (TCF) of natural gas cumulatively since the late 1960s, the U.S. could realistically produce at least that much additional gas, and possibly far more. Both the U.S. Energy Information Administration and the nonprofit Potential Gas Committee have recently estimated the total supply — including proven and potential reserves — at more than 2,400 TCF. But supply and reserve estimates are moving targets that vary based on factors such as geologic and technological constraints that limit our physical access to gas, and, even more critically, economic influences on natural gas pricing and demand.

Understanding of the geological underpinnings and distribution of natural gas resources, including conventional oil and gas deposits, tight gas and shale bed deposits, and offshore resources has accelerated greatly in the last decade, as have developments with the equipment and expertise used in extracting gas. In the years since the shale gas boom began, scientists and engineers have been using fine-scale methods like light microscopy and cathodoluminescence to image shale rocks in greater detail than ever before. Among their findings, they've gained more nuanced views of how the particular sizes and types of mineral grains, as well as the distribution of microscopic pores among the grains, affect shale's tendency to release stored methane during fracking. In addition, they've begun to grasp how and why productivity can vary greatly even within a single shale gas field. This has led in turn to better anticipation of areas likely to be more productive. Corresponding technological advancements

© Shutterstock.com/DabartiCGI

Cooperation in Colorado

In Colorado, as in other states where hydraulic fracturing ("fracking") has been increasingly used in oil and gas development, there has been a growing debate between both supportive and skeptical groups and individuals. Supporters see fracking as a proven and effective technique that spurred the domestic energy boom and its resultant economic benefits. Detractors are concerned about the potentially detrimental impacts of increased oil and gas production on the environment, public health and property values.

To de-escalate public feuds and advance the conversation over oil and gas development in Colorado, groups on both sides of the issue have been working to craft agreements that produce positive and tangible results for all involved.

In February 2014, for example, the state's Air Quality Control Commission approved landmark rules regulating methane emissions and leaks from oil and gas production sites — the first of their kind of in the country — after three of the state's largest energy producers, Anadarko Petroleum Corporation, Encana Corporation and Noble Energy, came together with the Environmental Defense Fund to develop draft regulations. Later that year, the Governor formed a 21-member task force to recommend how to use regulations to "balance land-use issues in a way that minimizes conflicts while protecting communities and allowing reasonable access to private mineral rights." The task force comprises representatives from local governments, civic groups, environmental groups, the oil and gas industry, agriculture and the home-building industry. The formation of this task force led to the withdrawal of four separate oil and gas drilling ballot initiatives, both against and supportive of drilling, from the November 2014 ballot. The withdrawal of the ballot initiatives avoided what were expected to be some of the most expensive and contentious campaigns in the state's history.

Steve Sonnenberg, professor and Charles Boettcher Distinguished Chair in Petroleum Geology at the Colorado School of Mines in Golden and a member of the state's oil and gas regulatory body, the Colorado Oil and Gas Conservation Commission, from 1997 to 2003, was asked to comment on these recent developments.

What is different about the current conversation over oil and gas regulation in Colorado that we have not seen before?

SS: We are seeing a new age in oil and gas development where the oil and gas companies, local governments and citizens are actually trying to work together in a cooperative fashion. These efforts have already facilitated progress through compromise.

What has been the public's response to seeing opposing groups come together to find middle ground on oil and gas production activities in Colorado?

SS: People have been very supportive. It promised to be a very difficult fall with those initiatives on the ballot in November 2014, and overall the citizens of Colorado supported the Governor in heading that off. There will always be some people out there who don't like the outcomes. But anything that keeps ballot initiatives from happening, especially when you have so many of them for and against an issue like oil and gas drilling, is probably a good thing.

Do you foresee Colorado setting a precedent for other states?

SS: Absolutely. People elsewhere will see that these activities can be successful in Colorado, and that Colorado can be a model for other states. The key is cooperation among affected groups to produce reasonable compromise agreements. •

have, for example, allowed multiple hydraulically fractured wells to be drilled from a single well pad and decreased the volumes of fluids used in the process, both of which lessen negative impacts on the surface. Such improvements have historically been implemented not just when mandated by law or regulation, but often voluntarily, especially when they increase

© Shutterstock.com/ixpert

© Shutterstock.com/Norgal

efficiency and decrease risk, to operators and the environment alike.

Historically, forecasts more than a few years into the future of the economics related to energy broadly, and natural gas specifically, have often been woefully inaccurate; there are simply too many moving parts. Expectations just a decade ago were for the U.S. to need to build multiple liquefied natural gas (LNG) import terminals to offset our dwindling production, yet no one foresaw the dramatic production boom leading to the country now converting import terminals for export. Because of uncertainties in many controlling factors, which can substantially enlarge or shrink supply, there is no unanimous agreement on just how long our abundant natural gas supplies could actually last. Could they last 100 years or longer, or just a few decades? Is that

sufficient to bridge our economy for energy as new alternative sources of energy come online? Is investment in a methane economy worth the effort? Will natural gas continue to offset coal production, and are there other parts of the economy for which it will generate new demand? Ultimately, how long our supply might last is based mostly on demand, which, in developed countries like the U.S., comes down to how much it costs and whether the supporting infrastructure exists.

Consumer Demand and Access

Natural gas consumption in the U.S. has been on the rise since the mid-2000s, growing from about 22 TCF in 2006 to 26 TCF in 2013. The vast majority of current consumption occurs in four sectors according to the U.S. Energy Information Administration: electricity generation (31 percent of consumption in 2013), industrial (28 percent), residential (19 percent) and commercial (13 percent). Electric power is the largest end-user of natural gas, and also the fastest-growing, primarily due to the ongoing replacement of coal-fired power plants with natural gas-burning plants. From 2000 to 2013, coal-generated electricity dropped from 52 percent of the total to 39 percent while

© Shutterstock.com/Lledo

India and other developing nations. This global demand combined with the shale gas boom is creating export opportunities for the U.S., particularly for LNG. Whereas U.S. imports of LNG peaked just eight years ago in 2007, construction is now underway to convert one former LNG import facility to an export facility and similar conversions have been approved

> Mudrocks are very fine-grained. They are a technical challenge to understand, and nobody predicted that we would be able to extract resources from these kinds of rocks.
>
> – Kitty Milliken

gas-generated electricity rose from 16 to 27 percent. This domestic growth is projected by most experts to continue, with natural gas soon supplanting coal as the leading source for electricity.

Demand growth in the domestic industrial, residential and commercial sectors is likely to be driven largely by the price of natural gas; it has been suggested that cheap gas could lead to a period of reindustrialization in the U.S., which could further drive demand.

The transportation sector is another potential driver for natural gas demand in the future, although this remains a significant unknown. Currently, use as a vehicle fuel accounts for less than 1 percent of natural gas consumption, as the American automotive market remains heavily dominated by gasoline. Making a dent in this oil-driven market could take years. But innovation in transportation is ongoing, and even a small impact on the gasoline economy — in which sales reach into the hundreds of billions of dollars annually in the U.S. — could be motivation enough for manufacturers to keep advancing natural-gas-fueled technologies and to develop a more extensive distribution system.

Outside the U.S., demand for natural gas has been rising steadily in recent decades and is projected to continue doing so, spurred by growth in China,

for three other facilities by the Federal Energy Regulatory Commission. The total LNG export capacity from these four facilities is expected to be about 7 billion cubic feet per day.

Consumer demand relies on the fundamental availability of supplies — in this case, accessible natural gas deposits. But in order for demand to grow, these supplies must also be made accessible to and usable for existing and prospective consumer end-uses. This will in turn require legal and regulatory policies that promote or, at the very least, allow natural gas to be transported nationally for use locally across the country. Additionally, national, state and local policies must coalesce to some extent to allow natural gas to be used on a broad scale, because opposing policies in adjacent states or municipalities can raise barriers to widespread distribution and utilization.

Improving Health, Safety, and Environmental Understanding

© Shutterstock.com/Budimir Jevtic

As with all energy sources, extraction, processing and use of natural gas comes with potentially significant hazards to the health and safety of people and the environment. In recent years, significant concerns — voiced largely in regard to hydraulic fracturing operations — have centered mostly on potential contamination of drinking water supplies and surface waters; water usage and sourcing; greenhouse gas emissions, including leaked, or fugitive, methane; land use and landscape degradation; and earthquakes caused by natural gas operations. Additional concerns have included potential effects of land and water use on animals, such as habitat fragmentation; environmental impacts from sand mining and other adjunct, fracking-related industries; increases in noise, traffic and traffic accidents near gas operations; and other public health impacts.

The tremendously rapid growth of natural gas production has spurred new studies to look at the validity of these concerns, and the results of these studies, which take time to thoroughly conduct and review, are only now beginning to make their way from researchers to the public. However, the time gap between the rise in public awareness and the release of such studies addressing potential problems has left a vacuum of reliable information, leading to under-informed statements and extrapolations about the risks, or lack thereof, of natural gas operations by groups on various sides of the issue — including industry, environmental groups, politicians and others. It has also complicated governmental efforts to regulate such operations. Collectively, the situation has fueled misconceptions, mistrust and strong emotional responses among the public about natural gas.

Results from scientific studies examining the impacts of fracking on groundwater and on induced earthquakes have, so far, seemed to suggest that, when done correctly and with adequate precaution, extracting natural gas from belowground is not inherently or unduly dangerous. But no individual study offers a complete answer or picture and there is still so much we don't know, meaning far more work — examining issues of geology, geography, hydrology, ecology, public health, epidemiology and other fields related to natural gas — is needed.

With additional study and understanding of the environmental and human health impacts, pragmatic and science-based regulation can be crafted to minimize accidents and impacts

> If you're going to ask communities to undertake the burdens associated with natural gas production, there must be reasonable assurances that industry and government are going to make sure that those burdens aren't wasted.
>
> – Mark Brownstein,
> U.S. Climate & Energy Program,
> Environmental Defense Fund

while allowing natural gas producers to operate efficiently. Such regulation may be most effective if it creates incentives for companies to develop cultures of accountability in their dealings at the state and federal levels as well as — vitally — within the communities in whose backyards they're operating.

To their credit, many companies in the energy industry already maintain internal programs and standards focused on best practices regarding the health, safety and environmental (HSE) impacts of their operations. Nonetheless, as in any complex technical endeavor — from manufacturing assembly lines to space exploration — unforeseen issues arise and lapses, both mechanical and human, occur. Accidents and surface spills related to gas operations have happened and will undoubtedly happen in the future. The key for long-term success is to combine sound regulation with an industry committed to addressing problems swiftly, adequately and transparently when they arise, and to learning and incorporating knowledge gained from their own and others' mishaps, and from community input, to continuously improve HSE practices.

Methane as a Fuel for Economic Growth

As with every potential future energy resource, there are both enablers of and barriers to the expansion of natural gas in the U.S. energy economy. Taking realistic stock of these factors is crucial in assessing whether natural gas could or should contribute substantially — perhaps even the majority — to our energy portfolio. The following are non-exhaustive lists of major enablers and barriers to the expanded use of natural gas for energy in the U.S. in the coming decades.

Enablers:
- Favorable price and economics of natural gas relative to other energy sources
- Continuous improvement in environmental, health and safety performance by industry
- Capacity for continued technological innovation
- Favorable geology holding substantial accessible reserves of natural gas
- Existing base of natural gas storage, processing and transport infrastructure
- Private ownership of land and mineral resource rights (which typically speeds decision making about resource development compared to public ownership)
- A central component of the energy mix needed to facilitate the transition to the energy sources of the next century

Barriers:
- Unfavorable price and economics of natural gas relative to other energy sources
- Difficulty in financing and expanding natural gas infrastructure (e.g., refineries, pipelines, LNG export ports) in a timely manner to meet demand
- Limited ability of technology, no matter how advanced, to increase gas recovery from difficult-to-access and low-yield resource plays
- Difficulty in comprehensively understanding factors that lead to productive shale gas reservoirs
- Political inability to make difficult, long-term decisions
- Inability of regulators to accommodate new technology in an adaptable and timely manner

Perhaps the most significant consideration, beyond the economic, technical and regulatory enablers and barriers listed, is the prevailing societal view of whether a reliance on natural gas is mostly positive or negative. The social license granted by consensus public opinion — at the national, state and local levels — can be either a substantial enabler or barrier, and its importance cannot be overstated.

The broad under-appreciation of the enormity of our energy requirements and of the ability of various energy sources to meet these requirements, combined with a shortage of science-based information on the potential impacts of natural gas extraction, muddies the waters for making clear, well-informed judgments. And inadequate, unclear and sometimes untruthful communication by stakeholder groups on all sides has hamstrung the honest, respectful and meaningful conversations about natural gas that could help provide the social license needed to decide its part in the mix of future energy sources in the U.S.

The social license granted by consensus public opinion — at the national, state and local levels — can be either a substantial enabler or barrier, and its importance cannot be overstated.

© Shutterstock.com/Shawn Hempel

The Path Forward

There is no single solution for how to best power our homes, our cars, our industries and our country, and no one group holds all the answers. Based on local conditions and resources, what is optimal for some communities may not be so for others. On the other hand, there are public policy decisions — about large-scale investments in energy science, technology and infrastructure, for example — that are best made at state and federal levels. For these reasons, conversations about energy need to happen at all levels of society and involve a variety of groups and voices — from energy producers, environmental groups and other nongovernmental organizations to scientists, economists, politicians and policymakers, landowners, local residents and others.

These conversations are the basis for the well-informed and critical decisions we must make about our future energy use. Productive paths forward begin with respectful engagement and dialogue, open recognition of the differing values and interests of diverse communities, and an honest accounting of the scientific, technological, economic and environmental realities with which we're confronted.

For decades, the notion of seeking such common ground over energy has been the exception rather than the rule. Today, with far greater public awareness of energy issues, things are different. And each of these groups can and must do better to engage, educate and listen.

Unfortunately, though unintentionally, the rapid advances in technology and geologic knowledge enabling the natural gas boom have vastly outpaced our understanding of its social and economic implications in the shifting energy landscape, leading to heated debates over energy development and slowing reasoned, effective policymaking. Only through active participation by all stakeholders can these social implications be fully understood in light of this rapid change.

Scientists can continue to improve our understanding of the sources, development and impacts of energy resources, and find articulate representatives to share their findings clearly.

Scientific societies, like AGI, can work to aggregate, condense and communicate the wealth of scientific and technological knowledge impartially and in terms that are accessible, respectful and meaningful to different groups. Politicians and regulators can take care to heed all sides in energy discussions and strive to separate reality and fact from fiction and rhetoric in crafting rules and policy. Companies can seek out social license for their actions by clearly communicating their intentions, addressing and respecting local values, interests and concerns, and holding themselves and any contractors who operate on their behalf accountable to the communities in which they're working. And as individuals, if we want a voice in the conversations, we have responsibilities to recognize the massive role of and need for energy in our lives and in the country, to educate ourselves about different energy resources, and to engage and listen to opinions other than our own.

With these guidelines as starting points, diverse stakeholders can find the needed common ground on which to build compromise and progress. •

© Shutterstock.com/Andresr

Informal Poll of Participants

During the meeting, participants were invited to respond to a series of informal poll questions. The results of these polls are a snapshot of some of the participants' opinions at that time.

Identify the 2 greatest social and political challenges that a methane economy faces

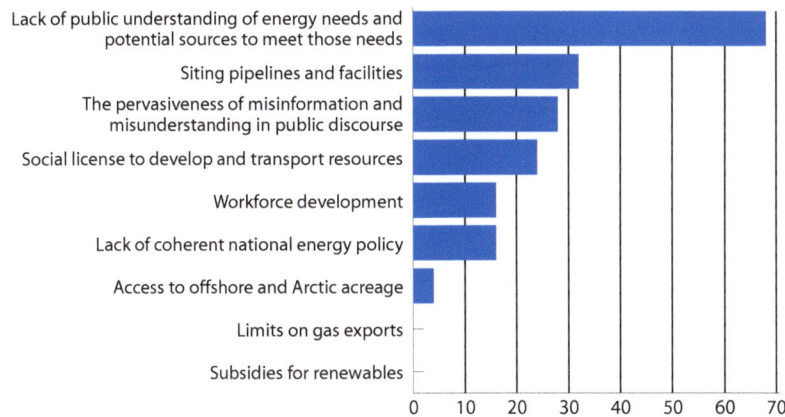

Percentage of responses from 25 respondents who each could choose 2 topics.

Identify the 2 greatest technical challenges that a methane economy faces

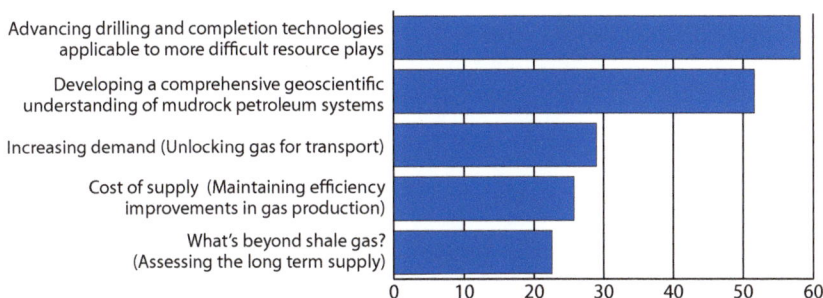

Percentage of responses from 31 respondents who each could choose 2 topics.

Identify the 2 principal enablers to achieving a methane economy

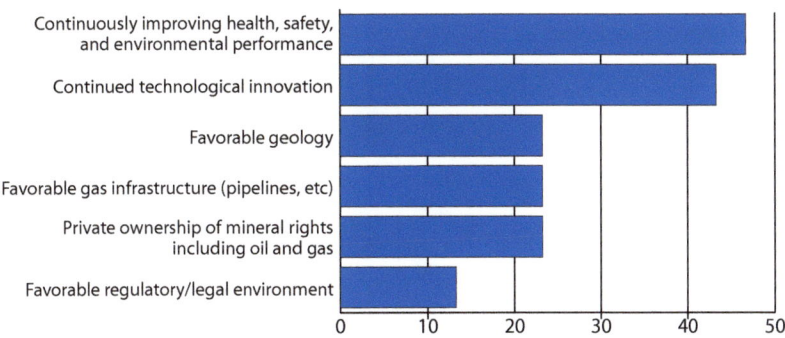

Percentage of responses from 30 respondents who each could choose 2 topics.

Keynote Speaker
Scott W. Tinker, Director, Bureau of Economic Geology, The University of Texas at Austin, and State Geologist of Texas
"Natural Gas: Fortune or Folly"

Scott W. Tinker is Director of the Bureau of Economic Geology, the State Geologist of Texas, a professor holding the Allday Endowed Chair and acting Associate Dean of Research in the Jackson School of Geosciences at The University of Texas at Austin, and Director of the Advanced Energy Consortium (AEC). He spent 17 years in the oil and gas industry prior to joining UT in 2000. Scott is past President of the American Association of Petroleum Geologists (AAPG), the Association of American State Geologists, and the Gulf Coast Association of Geological Societies. He has been a Distinguished Lecturer for the AAPG and Society of Petroleum Engineers, a Distinguished Ethics Lecturer for the AAPG, and the Geological Society of America (GSA) Halbouty Distinguished Lecturer.

Keynote Speaker
Katherine Lorenz, President and Treasurer, The Cynthia & George Mitchell Foundation
"George P. Mitchell: The Power of Individuals to Change the World"

Katherine was elected president and treasurer of the Cynthia and George Mitchell Foundation in January 2011. In late 2012, *Forbes Magazine* named Katherine "Ones to Watch" as an up-and-coming face in philanthropy.

Katherine serves on the board of directors of the Environmental Defense Fund, The Philanthropy Workshop (chair), Puente a la Salud Comunitaria, the Endowment for Regional Sustainability Science, Exponent Philanthropy, and the Amaranth Institute.

Katherine is a member of the Global Philanthropists Circle of the Synergos Institute, and sits on the Council on Foundations Committee on Family Philanthropy. She also serves on the National Academies' Roundtable of Science and Technology for Sustainability.

Katherine formerly worked as Deputy Director for the Institute for Philanthropy, whose mission is to increase effective philanthropy in the United Kingdom and internationally.

Prior to that, Katherine lived in Oaxaca, Mexico for almost six years where she co-founded Puente a la Salud Comunitaria, a non-profit organization working to advance food sovereignty in rural Oaxaca state through the integration of amaranth into the diet.

Before founding Puente, she spent two summers living in rural villages in Latin America with the volunteer program Amigos de las Américas and later served on their program committee and as a trustee of the Foundation for Amigos de las Americas.

Katherine is a frequent guest speaker on topics related to environmental sustainability, next generation philanthropy, and non-profit leadership. She holds a B.A. in Economics and Spanish from Davidson College.

Eyal Aronoff, Co-Founder, Fuel Freedom Foundation

Eyal Aronoff was a co-founder of Quest Software, which was sold to Dell for $2.4 billion in 2012. After leaving Quest in 2003, Eyal has started several successful companies in a variety of industries. His current focus is energy, algorithmic trading, and autism. Eyal is one of the largest funders of the effort to break the US oil addiction through the foundation he co-founded called Fuel Freedom Foundation. The Fuel Freedom Foundation goal is to break the oil addiction by opening the fuel market to competition both at the dealership and at the pump. Eyal is the producer of PUMP (www.PUMPTheMovie.com), a documentary movie that will forever change your attitude about fuel. Eyal is also one of the largest funders of clinical trials for treatments for autism and a supporter of the autism therapy portal Mendability.com, an affordable, home-based, sensory enrichment therapy.

Jesse H. Ausubel, Director, Program for the Human Environment, The Rockefeller University

Jesse Huntley Ausubel is Director of the Program for the Human Environment at The Rockefeller University in New York City. The program elaborates the technical vision of a large, prosperous society that emits little or nothing harmful and spares large amounts of land and sea for nature. Mr. Ausubel was a main organizer of the first U.N. World Climate Conference in 1979. He helped develop the concept of "decarbonization" and published the first paper using the word in 1991. From 2006-2010, he served as a director of the Electric Power Research Institute and now serves on its Advisory Council.

Lawrence Bengal, Director, Arkansas Oil and Gas Commission

Lawrence Bengal holds a degree in Geology from the University of Wisconsin and has over 35 years experience in the public and private sectors. Mr. Bengal currently serves as Director of the Arkansas Oil and Gas Commission and as a Commissioner on the Pollution Control and Ecology Commission.

Mr. Bengal has served as the Governor's representative for Illinois and currently serves as the Governor's representative for Arkansas to the Interstate Oil and Gas Compact Commission (IOGCC), where he has served as IOGCC Commission Vice-Chair and Chair of the Environmental Committee and currently serves as Chair of the IOGCC Carbon Capture and Geologic Storage Task Force and Chair of the IOGCC-GWPC State Oil and Gas Regulatory Exchange.

Mark Brownstein, Associate Vice President & Chief Counsel, U.S. Climate & Energy Program, Environmental Defense Fund

Mark Brownstein is Associate Vice President and Chief Counsel of the U.S. Climate and Energy Program at Environmental Defense Fund (EDF). Mark leads EDF's team on natural gas development and delivery. In addition, he specializes in a variety of utility-related issues including electric grid development and wholesale and retail market design.

Prior to joining EDF, Mark held a variety of business strategy and environmental management positions within Public Service Enterprise Group (PSEG), one of the largest electric and gas utility holding companies in the United States.

Mark's career includes time as an attorney in private environmental practice, an air quality regulator with the New Jersey Department of Environmental Protection, and an aide to a member of the U.S. House of Representatives.

Mark is a member of the Electric Power Research Institute's Public Advisory Committee.

Mark holds a J.D. from the University of Michigan Law School, and a B.A. from Vassar College.

John B. Curtis, Professor Emeritus of Geology and Geological Engineering, Director, Potential Gas Agency, Colorado School of Mines

John B. Curtis is Professor Emeritus of Geology and Geological Engineering and Director, Potential Gas Agency, at the Colorado School of Mines. Dr. Curtis has been at the Colorado School of Mines since July 1990. He had 15 years prior experience in the petroleum industry with Texaco, Inc., SAIC, Columbia Gas, and Brown & Ruth Laboratories/Baker-Hughes. He serves on and has chaired several professional society and natural gas industry committees, which previously included the Supply Panel, Research Coordination Council, and the Science and Technology Committee of the Gas Technology Institute (Gas Research Institute). He co-chaired the American Association of Petroleum Geologists (AAPG) Committee on Unconventional Petroleum Systems from 1999-2004 and is an invited member of the AAPG Committee on Resource Evaluation. He was a Counselor to the Rocky Mountain Association of Geologists from 2002-2004.

Wendy Harrison, Professor of Geology and Geological Engineering, Colorado School of Mines

Wendy J. Harrison is a tenured Professor of Geology and Geological Engineering at Colorado School of Mines. Her fields of scholarly expertise are in geochemistry and hydrology as well as geoscience education and she has published papers in topics that range from impact shock metamorphism in lunar materials, the formation of gas hydrates and their role in CO_2 sequestration, metals uptake by trees in mined lands, and mitigating respiratory quartz dust hazard. During her career in academia at Colorado School of Mines, she has served as Director of the McBride Honors Program in Public Affairs for Engineers, and Associate Provost and Dean of Undergraduate Studies and Faculty. Dr. Harrison recently completed an appointment at the National Science Foundation as Division Director for Earth Sciences in the Geosciences Directorate. She currently serves as an advisor to the Petroleum Institute, Abu Dhabi and Nazarbayev University, Kazakhstan, in the foundation of in-country research and education programs in earth resources. Educated at the University of Manchester, UK, she held a pre-doctoral fellowship at The Geophysical Laboratory of the Carnegie Institution of Washington and a National Research Council research fellowship at NASA-Johnson Space Center. Her work experience includes 8 years as a senior research geologist for Exxon Production Research Company in Houston, Texas.

Doug Jordan, Director, HS&E Corporate Environmental Programs, V+ Development Solutions Division, Southwestern Energy Company

Doug Jordan is currently Director, HS&E Corporate Environmental Programs, V+ Development Solutions Division, a division of Southwestern Energy Company. The mission of V+ Development Solutions is to identify, develop, and implement solutions to the challenges of unconventional resource development that strike an appropriate balance among the environmental, social, and economic impacts of the Company's activities. He has been with Southwestern Energy since 2009.

Mr. Jordan has over 28-years of HSE experience including experience as a regulatory agent, consultant, and industry professional. His industry experience is predominately oil and gas oriented in the production, gathering and processing, and transmission and storage sectors in over 30 states. He is actively engaged in industry trade associations and currently serves as Chair of the Environmental Committee for the Gas Processor Association. He is also currently engaged in the Technical Work Groups associated with several methane measurement and monitoring initiatives. Mr. Jordan graduated from Oklahoma State University in 1985.

Alan Krupnick, Senior Fellow and Director, Center for Energy Economics and Policy, Resources for the Future

Alan Krupnick is a Senior Fellow and Director of the Center for Energy Economics and Policy (CEEP) at Resources for the Future (RFF). Krupnick's research focuses on analyzing environmental and energy issues, in particular, the benefits, costs, and design of pollution and energy policies, both in the United States and in developing countries, with an emphasis on China. As head of CEEP, he leads RFF's research on the risks, regulation, and economics associated with shale gas development and has developed a portfolio of research on issues surrounding this newly plentiful fuel.

David Levinson, Department of Civil, Environmental, and Geo- Engineering, University of Minnesota

David Matthew Levinson is an American civil engineer and transportation analyst, currently a professor at the University of Minnesota, where he holds the RP Braun/CTS Chair in Transportation. He has authored or co-authored 4 books, edited 3 collected volumes, and authored or co-authored over 100 peer-reviewed articles on various aspects of transportation. He is a founder of the World Society for Transport and Land Use Research. In 1995 he was awarded the Charles Tiebout prize

in Regional Science by the Western Regional Science Association and in 2004, the CUTC-ARTBA New Faculty Award. His travel behavior research was featured in the book *Traffic* by Tom Vanderbilt.

Richard Liroff, Founder and Executive Director, Investor Environmental Health Network

Dr. Richard Liroff is founder and Executive Director of the Investor Environmental Health Network (www.iehn.org). He earned a Ph.D. in Political Science from Northwestern University and a B.A. in Politics from Brandeis University.

Since 2009 Dr. Liroff has led investor efforts to promote increased disclosure by energy companies on risks from horizontal drilling and hydraulic fracturing operations in "shale plays." He is principal author of *Extracting the Facts: An Investor Guide to Disclosing Risks From Hydraulic Fracturing Operations*. It identifies twelve core management goals, practices to implement them, and indicators for reporting progress. He is also lead author of *Disclosing the Facts*, a disclosure scorecard based on *Extracting the Facts*.

Kenneth Medlock, James A. Baker III and Susan G. Baker Fellow in Energy and Resource Economics, and Senior Director, Center for Energy Studies, James A. Baker III Institute for Public Policy, Rice University

Kenneth B. Medlock III, Ph.D., is the James A. Baker, III, and Susan G. Baker Fellow in Energy and Resource Economics at Rice University's Baker Institute and Senior Director of the Center for Energy Studies, as well as an adjunct professor and lecturer in the Department of Economics at Rice University. He is a principal in the development of the Rice World Natural Gas Trade Model, aimed at assessing the future of international natural gas trade. He has published numerous scholarly articles in his primary areas of interest: natural gas markets, energy commodity price relationships, gasoline markets, transportation, national oil company behavior, economic development and energy demand, and energy use and the environment.

Kitty Milliken, Senior Research Scientist, Bureau of Economic Geology, Jackson School of Geosciences, University of Texas at Austin

Kitty Milliken is a Senior Research Scientist at the Bureau of Economic Geology in the Jackson School of Geosciences at the University of Texas at Austin. She received degrees from Vanderbilt University (B.A.) and the University of Texas at Austin (M.A., Ph.D.). Her research concerns the integration of petrographic and analytical methods to decipher the chemical and mechanical histories of sedimentary rocks. Her current focus is on the fine grained sedimentary rocks that host unconventional reservoirs for oil and gas.

Richard Nehring, President, Nehring Associates

Richard Nehring has been President of Nehring Associates since he founded the company in 1983. During this period, he designed the Significant Oil and Gas Fields of the United States Database and its subsequent expansions and directed the initial development and subsequent updates, upgrades, and expansions of the database. Since the initial release of the database in 1985, Mr. Nehring has written more than 20 papers and presentations using the database. Since 1980, Mr. Nehring has served on numerous professional and scientific committees dealing with oil and gas resource and supply issues, including three National Petroleum Council task groups and four National Research Council Committees. He has been a member of AAPG's Committee on Resource Evaluation since its founding in 1993 and is chairman of this committee from 2011 to 2014. He was also Chairman and Organizer of the AAPG Hedberg Research Conference on Understanding World Oil Resources in November 2006. Prior to founding Nehring Associates, Mr. Nehring was project director of fossil fuel supply issues for the Energy Policy Program of the Rand Corporation for ten years. His major studies during the period covered giant oil fields and world oil resources, the discovery history and size distribution of U.S. oil and gas fields and their implications for ultimate resources, the heavy oil resources of the United States, and Mexico's petroleum and U.S. policy.

L. Renee Orr, Chief, Office of Strategic Resources, Bureau of Ocean Energy Management, U.S. Department of the Interior

Renee Orr has more than 25 years of experience with the Department of the Interior. She is a senior executive on the Bureau of Ocean Energy Management leadership team.

As the Chief of the Office of Strategic Resources, Ms. Orr oversees development and implementation of the Nation's offshore oil and gas and marine mineral leasing programs. She also oversees the assessment of offshore oil and gas resources as well as ensuring that the Nation receives fair market value for these valuable assets. She completed the Department of the Interior's Senior Executive Service Candidate Development Program in 2001.

David Pursell, Managing Director, Head of Securities, Tudor, Pickering, Holt & Co.

Dave Pursell serves as Managing Director and Head of Securities at Tudor, Pickering, Holt & Co. (TPH). Dave is responsible for TPH's analysis of global oil & gas markets, including inventory and price forecasts, supply/demand modeling and rig count/production relationships. He was past Chairman of the IPAA Supply Committee and sits on the Investment Committee of TPH Partners LP, TPH's private equity division. Dave is a board member of private energy companies Oxane Materials and Unconventional Gas Resources. He was a Founding Partner of Pickering Energy Partners, the predecessor to TPH. Prior to that, he was Director of Upstream Research at Simmons & Company, International and spent eight years as manager of petrophysics at S.A. Holditch & Associates, now a division of Schlumberger. He gained operational experience with ARCO Alaska, Inc., conducting field engineering and operations. He holds a B.S. and M.S. in Petroleum Engineering from Texas A&M University.

Randy Randolph, Vice President, Southern Gas Association

On December 1, 2008, Randy joined the Southern Gas Association as a vice president. After retiring from Cinergy in 2005, Randy formed Double R Associates and began providing energy management advisory services. From 1997 to 2005, Randy worked in various executive capacities at Cinergy Corp but most recently as VP, Gas Distribution Operations. In 1995-97, Randy provided independent energy management and natural gas supply and marketing consulting services. From 1993 to 1995, he served as VP Gas Resources for Transok and was responsible for gas acquisition, marketing, transportation and energy risk management services. Over 17 years with The Williams Companies, he served in many capacities including President of Williams Energy. During those years he directed the operation of gas pipelines, marketing, natural gas liquids and energy trading. Randy received BS and BBA degrees from the University of Texas and completed the Advanced Management Program at Harvard Business School.

Berry H. (Nick) Tew, Jr., State Geologist of Alabama and Oil and Gas Supervisor, Geological Survey of Alabama and State Oil and Gas Board of Alabama

Dr. Nick Tew has served as Alabama's State Geologist and Oil and Gas Supervisor since 2002. In these capacities, he directs the Geological Survey of Alabama and the staff of the State Oil and Gas Board of Alabama. Nick previously served as President of the American Geosciences Institute, President of the Association of American State Geologists, Vice-Chairman of the Interstate Oil and Gas Compact Commission, and Chairman of the U.S. Department of the Interior Outer Continental Shelf Policy Committee. He also serves on the National Petroleum Council and is a Fellow in the Geological Society of America.

connecting earth, science, and people

About AGI: AGI was founded in 1948, under a directive of the National Academy of Sciences, as a network of associations representing geoscientists with a diverse array of skills and knowledge of our planet. The Institute provides information services to geoscientists, serves as a voice of shared interests in our profession, plays a major role in strengthening geoscience education, and strives to increase public awareness of the vital role the geosciences play in society's use of resources, resilience to natural hazards, and the health of the environment.

AGI connects Earth, science, and people by serving as a unifying force for the geoscience community. With a network of 50 member societies, AGI represents more than a quarter-million geoscientists. No matter your individual discipline, AGI's essential programs and services will strengthen your connection to the geosciences.

EARTH Magazine: This monthly publication explores the science behind the headlines. EARTH magazine gives readers definitive coverage on topics from natural resources, energy, natural disasters and the environment to space exploration and paleontology and much more.

Education and Outreach: AGI Education offers products and services for K-12 educators, including NSF-funded curricula, high-definition videos, classroom activities, teacher professional development, and online resources.

GeoRef: GeoRef is a comprehensive, bibliographic database containing over 3.5 million references to geoscience journal articles, books, maps, conference papers, reports and theses.

Policy and Critical Issues: Geoscience Policy works with AGI member societies and policy makers to provide a focused voice for the shared interests of the geoscience profession in the federal policy process. Critical Issues provides a portal to comprehensive, impartial geoscience information for decision makers.

Workforce: AGI produces the Directory of Geoscience Departments publication on human resources of the U.S. geosciences community. It collects data on the supply and demand of geoscientists, and works with other organizations and government agencies to ensure that the health of the profession is understood.

Earth Science Week: Reaching over 50 million people a year, Earth Science Week promotes awareness of Earth science and appreciation of the geosciences' role in society. This international public awareness campaign, organized each October by AGI, provides informational resources, educational materials, and a variety of events and activities for students, teachers, and others. Program partners in government, industry, and the nonprofit sector unite to advance these efforts and continue the solid track record of success of this nearly two-decade-old initiative (www.earthsciweek.org).

american geosciences institute

Center for Geoscience & Society: The Center links geoscience information to diverse, non-specialist audiences, with a particular emphasis on communicating with decision makers at all levels and with educators in non-geoscience disciplines.

AGI Foundation: The Foundation is the principal source of U.S. tax-deductible endowment and programmatic contributions to the American Geosciences Institute from industry, private foundations, and individual donors.

The Critical Issues Program is a new program at the American Geosciences Institute. Its main purpose is to make geoscience information more discoverable to decision makers at all levels.

Critical Issues Website

www.americangeosciences.org/critical-issues

The Critical Issues website is a hub for decision-relevant, impartial geoscience information on many of society's most pressing issues. The Critical Issues website aggregates information from multiple geoscience organizations, making it easy for users to find trusted, comprehensive information from across the geosciences at one location.

The website's topic pages highlight resources from the geoscience community on climate, energy, hazards, mineral resources, and water, with easy-to-digest summaries, answers to common questions, portfolios of maps and tools, and links to more detailed documents about the issue.

Critical Issues Research Database

www.americangeosciences.org/critical-issues/research-database

The Critical Issues Research Database allows users to quickly search for topics, and link through to the documents on the websites of the organizations that produced the content.

- Contains more than 3,000 factsheets, reports, position statements, and case studies; expanding weekly
- Decision-relevant geoscience information, indexed for legislative staff and researchers
- Impartial sources:
  - State geological surveys (70% of documents)
  - U.S. Geological Survey (21%)
  - Geoscience and other organizations (9%)
- Links users to the original source of the documents
- Searchable by location

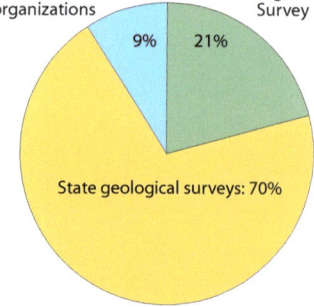

Critical Issues Research Database document sources by percent

Geoscience and other organizations 9%

U.S. Geological Survey 21%

State geological surveys: 70%

Critical Issues GeoIssues Webinars

Coming in 2015!

The Critical Issues program will be launching its GeoIssues webinar series in 2015 to bring geoscientists and decision makers together to discuss potential solutions to these challenges.

@AGI_GeoIssues

Center for Geoscience & Society

american geosciences institute

american geosciences institute

connecting earth, science, and people

www.ingramcontent.com/pod-product-compliance
Lightning Source LLC
Chambersburg PA
CBHW041615180526
45159CB00002BC/868

* 9 7 8 1 5 0 8 8 4 3 5 0 4 *